Bibliographic information published by the German National Library:

The German National Library lists this publication in the National Bibliography; detailed bibliographic data are available on the Internet at http://dnb.dnb.de .

Imprint:

Print and binding: Books on Demand GmbH, Norderstedt Germany
ISBN: 9783668548633

This book at GRIN:

http://www.grin.com/en/e-book/377268/the-effects-of-monetary-policy-in-the-us-the-vector-error-correction-model

Colin Tissen, E Voisin

The Effects of Monetary Policy in the US. The Vector Error Correction Model (VECM) compared to the Structural Autoregressive Model (SVAR)

GRIN Publishing

Empirical Econometrics 2
Evaluation of monetary policy in the US

C. Tissen E. Voisin

June 29, 2017

Contents

1 Introduction

Estimating and anticipating the effects of monetary policies on the economy has always been one of the main concerns in macroeconomics. Most of monetary policies are systematic; namely, dictated by rules. However, economists are primarily interested in non-systematic movements in monetary policy in order to estimate their causal effects on macroeconomic variables. For this purpose, deviations from the monetary rules are used to investigate the relation between monetary policy and the economy through the simulation of shocks or their structural identification (Ramey, 2016).

This paper investigates the effects of monetary policy in the US by comparing a system of equations – estimated from a VECM (vector error correction model) – to a SVAR (structural autoregressive) model. Vector error-correction models are used when there exist long-run equilibrium relationships between non-stationary data integrated of the same order. Those models imply that the stationary transformations of the variables adapt to disequilibria between the non-stationary variables in the model. In contrast, SVAR models focus on the contemporaneous interdependence between the variables (Pfaff, 2008). We apply these two methods on a model with a contractionary monetary policy which affects the short-term interest rate. Following Sims and Zha (2006) we use a shock to the Treasury Bill rate instead of a shock to the Federal Funds rate, this is also in line with Sims (1986) and Amico & King (2017).

The paper continues as follows. First, a description of the data is given. Secondly, it presents a system of equations built from the LSE approach, aiming at macroeconomic simulations. Thirdly, it compares results obtained from the previous part to those obtained using SVAR impulse response functions (IRFs) identified with sign restrictions. The paper focuses on the impact of the simulated policies or monetary shocks on GDP and its growth rate.

2 Data

The data was retrieved from the FRED database. The data set consists of the following monthly time series from July 1959 to March 2017.

- LY: log of real GDP, seasonally adjusted
- $LRM2$: log of real M2, seasonally adjusted

- LRI: log of real gross private domestic investment, seasonally adjusted
- $LPCE$: log of real personal consumption expenditures, seasonally adjusted
- π: Inflation, computed from Consumer Price Index, seasonally adjusted
- R^m: nominal own return on M2
- R^b: interest rate on three-month Treasury Bills as a proxy for the Federal funds rate
- LTR: log of total reserves
- $UNRATE$: unemployment rate
- π_{commod}: The inflation of the commodity price index for all commodities.

The data for output and investment could only be found quarterly and were therefore interpolated linearly into monthly time series. Moreover, the two measures of inflation were computed as the growth rate of the consumer price index – or commodity price index – over one year, to be of the same scale as the interest rates.
According to augmented Dickey-Fuller (ADF) tests, all variables are integrated of order 1, with the exceptions of π_{commod}, which is stationary and $unrate$, which appears to behave as a long memory process. We will nonetheless consider the unemployment rate as stationary instead of taking a fractional difference, which would complicate the interpretation.

3 The LSE approach

For the creation of a system of equations, the LSE approach is used instead of the Cowles Commission (CC) approach. The CC methodology assumes that the structural form of the data generating process (DGP) is known qualitatively and the reduced form is then derived from it. The LSE approach explains the failure of the CC approach; there is a lack of attention for the statistical model underlying the econometric structure adopted. The LSE approach recognises that economic theory suggests the general specification of the relevant form but the precise DGP is almost never known. The reduced form takes a central role within this approach in that it represents the crucial probabilistic structure of the data. The approach is thus turned backwards, the reduced form will lead to the structural one.

3.1 Vector error correction model

Various studies (e.g., Favero (2001)) have indicated cointegration – long-run equilibrium relationships – between the macroeconomic variables used here. Hence, to determine the potential number of cointegrating equations between the variables, we first estimated the number of lags to be included based on a VAR in levels, including lags of π_{commod} and $unrate$, and a dummy representing the post-2008 crisis period as exogenous variables. The addition of commodity inflation will capture the effects of oil price crises or other significant commodity fluctuations that could have an effect on variables in our model. Moreover, after the 2008 crisis, requirements for banks and monetary approaches have drastically changed, hence, the inclusion of the dummy will take that into account. Nonetheless, in the period 1979-1982 the monetary regime of the FED changed, from a strategy aimed at controlling interest rates to a non-borrowed reserves targeting regime. The existence of regime shifts and associated structural breaks is seen to undermine traditional forecasting procedures severely, hence, the sample was reduced to 1983-2017 (411 data points). Schwartz information criterion (BIC) advocates the use of two lags in levels, thus, only one lag for the first differences in the VECM. Consequently, the Johansen cointegration test was performed which suggested four cointegrating relationships. Various restrictions based on economic theory were imposed on the coefficients in the cointegrating equations and it appeared that only three of them were statistically different. This can be due to the relatively large number of variables included in the models as well as the exogenous ones that alter the critical values of the test. The three cointegrating relationships are presented in Table 1.

The first equation (ECM1) represents the long-run relationship between output and private consumption. Consumption depends positively on output since people tend to consume more when they have a higher income. The second equation (ECM2) is the long-run relationship of the interest on money with the market rate and inflation. An increase in inflation will, in the long-run, decrease the short-term interest rate since investors will be more willing to save than to invest, banks will therefore borrow less from the Fed as their minimum total reserves would be attained. Such a decrease in borrowing will thus decrease the 'price of money', namely, the risk-free rate. On the other hand, an increase in market returns will increase investments and consequently increase borrowing from investors. Banks will thus

Cointegrating Eq:	CointEq1	CointEq2	CointEq3
LY(-1)	-1.445822 (0.05695) [-25.3875]	0.000000	0.000000
LRM2(-1)	0.000000	0.000000	-2.081233 (0.18134) [-11.4770]
LPCE(-1)	1.000000	0.000000	0.000000
LRI(-1)	0.000000	0.000000	0.000000
INFL(-1)	0.000000	0.545559 (0.10517) [5.18764]	0.000000
M2OWN(-1)	0.000000	-1.831961 (0.08702) [-21.0533]	0.000000
TBILL3(-1)	0.000000	1.000000	0.057964 (0.01155) [5.01892]
LTOTRES(-1)	0.000000	0.000000	1.000000
@TREND(59M07)	-0.001006 (0.00015) [-6.51031]	-0.009663 (0.00263) [-3.66806]	0.007259 (0.00064) [11.2965]
C	5.293332	4.367063	-1.422141

Table 1: Cointegrating equations

themselves need to borrow more from the Fed, leading to an increase of risk-free rate. The last equation (ECM3) is the long-run relationship between the total reserves with real M2 and short-term interest rate. As expected, if the amount of money available in the economy increases, the possibility for people to save expands as well, which in the long-run increases the total reserves of banks. However, we would expect *Tbill3* to have a positive effect on total reserves in the long-run since when it increases, people are more willing to save than to consume, intuitively leading to an increase in total reserves. However, as the money supply increases, market interest rate

($M2own$) decreases, which, in the long-run – from equation 2 – will lead to a decrease of $Tbill3$. Hence, the negative coefficient (and its small magnitude) might be due to the indirect effect of an increase in M2 on the interest rate. It can as well come from the fact that it decreases total reserves in the short-run since an increase in interest rate makes money more expensive for banks, who borrow less from the Fed, which decrease their total reserves in the short-term.
LRI and $Tbill3$ do not depend on the long-run equilibria, and can therefore be considered as exogenous in our model. Their equations are thus discarded and only the 6 other ones – additionally to an unemployment rate equation including the same regressors as in the VECM – remain to create a system of equations, in which contemporaneous growth rates of $Tbill3$ and investment are added. Tests of autocorrelation in the residuals were performed to check for missing dynamics and each equation was then adapted from the findings.

3.2 Simulations

The actual monetary policy (baseline) of the Fed after the 2008 crisis – keeping the interest rate low – is compared to an alternative policy scenario in which interest rates were kept at a relatively high rate of 4%, similar to that of the end of 2007. As mentioned before, we follow Sims (1986), Sims and Zha (2006), and Amico & King (2017), and use the three month Treasury Bill ($Tbill3$) as a proxy for the Federal funds rate. The Federal Funds rate is one of the most influential interest rates in the U.S. economy. It affects both monetary and financial conditions, which in turn impact the economy in general, through unemployment, growth or inflation for instance. The simulations were dynamically estimated – they consists in solving the system for the endogenous variables based on the exogenous ones – and the effects of the two different policies on the level of output are presented in Figure 1 below. The change in output is not substantial, and the simulations are, overall, lower than the actual GDP. What is striking is that in the long run, the policy of keeping interest high leads to higher output than the one with low rates. This does not make sense since when rates are high, consumers are less willing to borrow money, hence people tend to consume less as they cannot borrow, leading to a decrease in output. However, within the first year of the policy, high interest rates actually decreases output, which can be seen in Figure 2. The economy (depicted here with the model used) probably adapts to the high interest due to, for instance, the long run equilibria.

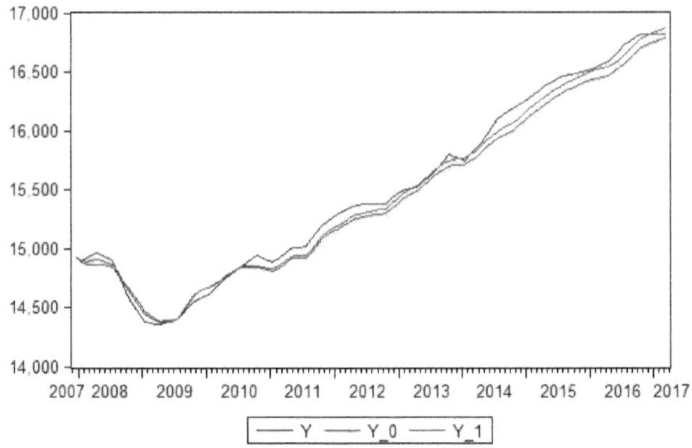

Figure 1: Effect of the 2 policies on level of output

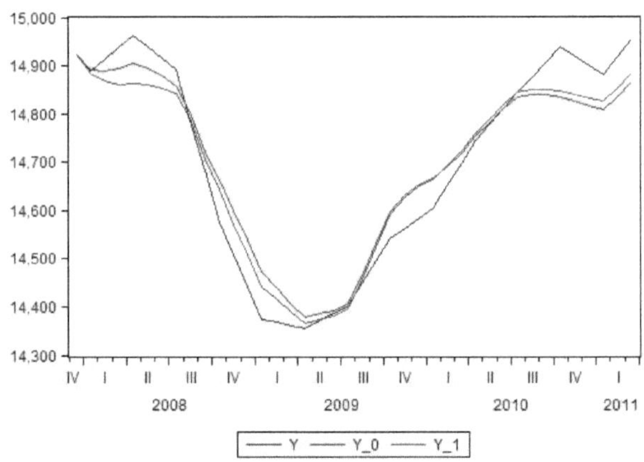

Figure 2: Effect of the 2 policies on level of output within the 3 first years

The long run adjustments can be seen in the growth rate of output as well (Figure 3). For a few months after the policies, the one with high interest rates leads to lower output growth than the baseline. However, this ordering is interchanged a few times until the two rates stabilize at the same level after approximately a year and a half. This phenomenon is more apparent in Figure 4, picturing only the three first years after the policies.

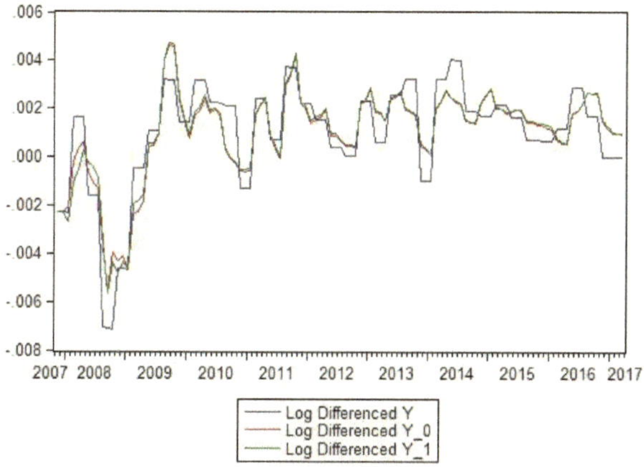

Figure 3: Effect of the 2 policies on output growth

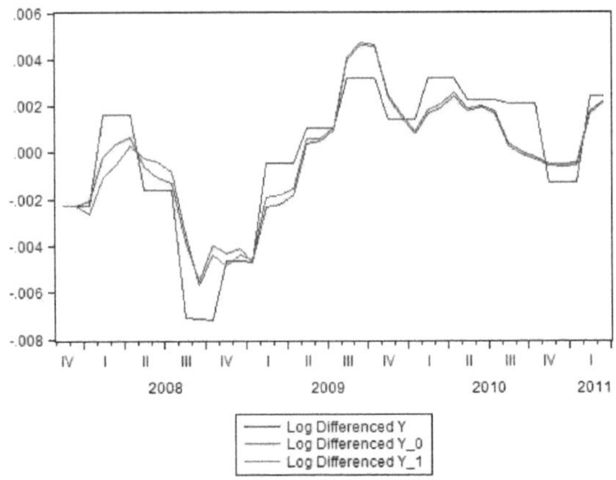

Figure 4: Effect of the 2 policies on output growth within the 3 first years

Hence, the results are surprising, yet, in the output regression, the coefficients of *Tbill3* are not significant. They are not significant either in the equations of the variables mostly influencing output (investment and consumption), and GDP does not react to the disequilibrium of ECM2 or ECM3 – in which the interest rate appear. This either indicates that output is relatively independent from the short-term interest rate – which is counter intuitive – or that the effects of *Tbill3* are wrongly estimated, possibly due to other variables absorbing its effect, or to instability of the coefficients. Concerning the other variables in the model, except $M2own$ in the scenario which significantly diverges from the the actual and baseline series, all other series do not seem to be that sensitive to the interest rate. Moreover, their growth rates, similarly to that of GDP, converge quickly to the same rate in both the baseline and the scenario. Performing static forecasts – using historical data for the lagged endogenous variables – would not have enabled us to investigate long-term effects of a policy since it consists of various one-step ahead forecasts and each point carries past information concerning actual data and not the simulated policy.

3.3 Coefficients instability

This paper focuses on monetary policies based on interest rates targeting, thus, the period in which the Fed regime shifted to non-borrowed reserves targeting could have significantly altered the results. Yet, the remaining sample, from 1983 onwards might not be free of breaks either, since it includes various crises. Tests of structural breaks were thus performed on the equations and the presence of two significant breaks was discovered, namely the dot-com bubble (early 2000s) and the 2008 crisis. The dummy included in our model might take most of the effects of the latest one into account, though the dot-com bubble was ignored. Nonetheless, the last one had more impacts on the monetary sector since, for instance, minimum total reserves imposed on banks went significantly up inducing an increase of total reserves by a factor of 19 within the last four months of 2008 (from 45 billion dollars to 860). As mentioned before, structural breaks might alter the cointegrating equations as well as the coefficients estimations, thus, further research could be done, by separating the whole sample into more subsamples and see whether the cointegration between the variables and their effects on the others are stable over time or not.

4 Structural VAR

Throughout history, researchers used several methods to evaluate the effects of monetary policy on the economy. In the 1970s and 1980s they used shocks to the money stock to measure monetary policy shocks. However, they then discovered that the importance of these shocks was significantly reduced when including interest rates in the VAR and the conclusion was drawn that monetary policy was not relevant for understanding economic fluctuations (Ramey, 2016). However, at the end of the 1980s, Romer and Romer (1997) counterclaimed the irrelevance of monetary policy by using a narrative series to define so-called "Romer Dates" at which the Federal Reserve used contractionary measures to decrease inflation. Their main result was that production decreased heavily after these Romer Dates. Yet, a few years later, their research was criticised from not separating the exogenous shocks from the endogenous monetary policy responses. In the following years, research focused more on the federal funds rate as an important indicator of monetary policy and emphasis was put on the specification of the monetary policy function (Ramey, 2016).
Alongside the narrative approach, several other methods for identification were developed; these include high frequency identification and sign re-

strictions for instance. As its name indicates, the first method uses high-frequency data, such as news announcements, to deal with the problem of foresight; it ensures that shocks are unanticipated. However, this shock might not be exogenous without further assumptions. For example, when examining monetary policy, private information of the Fed must be controlled for. Moreover, the sign restriction approach aims to correct for possibly biased identification results that are caused by a priori theorizing. Researches have "reasonable" results in mind and if those results are not achieved they are classified as a puzzle. Uhlig (2005) argues for a new approach, complementing others like Faust (1998), that explicitly uses these a priory theories. This is done by imposing, as little as possible, sign restrictions on the impulse responses, leaving the response of interest free.
We decided to use the sign restrictions approach for several reasons. First, using the narrative or high frequency method requires certain information that is not available to us, or that is not obtainable within the time-period of our research. Secondly, the sign restrictions approach has been successfully used to investigate monetary policy in the past (Faust, 1998; Uhlig, 2005; Gobbi & Willems, 2011; Castelnuovo et al., 2012). This provides us with enough sources to determine possible restrictions.

4.1 Sign restrictions

One of the most influential papers regarding sign restrictions was written by Faust (1998). He uses a 6-variable and a 13-variable model. His models contain – among others – all variables that we use, except that he uses M1 instead of M2. However, as argued by Lothian (2009) the effects on M2 will be comparable to those on M1. Another influential paper on sign restrictions applied to the same topic and similar data was written by Uhlig (2005). We hence mostly based our restrictions on those two papers, which were in accordance with Leeper, Sims, and Zha (1996) and Christiano, Eichenbaum, and Evans (1997). The response of *Tbill3* will be restricted to remain positive for 6 months after its own positive shock. Moreover, as mentioned before, *Tbill3* and *M2own* tends to move together, hence, it will be restricted to react positively to a contractionary monetary shock – an increase in the interest rate. The only other variable restricted to react positively, on impact, will be the unemployment rate.
These restrictions are justified in the following manner: higher short-term interest rates decrease the money supply (M2) and lead to a decrease in investments since loans for businesses are more expensive, this causes the demand for workers to fall and unemployment rises consequently. More-

over, consumption decreases which ultimately results in lower inflation rates. However, an issue with VARs on monetary policy is that a contractionary monetary policy shock often resulted in rising inflation in the short-run; this is called the "price puzzle" (Ramey, 2016). Boivin, Kiley, and Mishkin (2010) argue that the standard VAR results for inflation are highly sensitive to the specification. One way of dealing with this is incorporating commodity prices in the model; Sims (1992) showed that the inclusion of this variable significantly reduced the "price puzzle". However, it did not solve the problem in the model of Boivin et al. (2010). In order to fix it they included a measure of future inflation; they indeed found that this eliminated the price puzzle. Sims (1992) conjectured that policy shocks which are associated with substantial price puzzles are actually confounded with non-policy disturbances that signal future increases in prices. The "price puzzle" is therefore an anomaly in simpler VAR models due to information asymmetry between policy makers and analysts. Moreover, as argued before, contractionary policy makes it harder for banks to borrow from the Fed, resulting in a total reserves decrease in the short-term. Thus, consumption, total reserves, M2 and investment will be constrained to react negatively to the shock for 3 months. Moreover, two cases will be drawn, one in which inflation will as well be restricted to react negatively and one in which it will not be constrained.

4.2 Impulse response functions

Before identifying the shocks, a VAR – with stationary data – is required to be estimated. Once again, BIC serves as an indicator of the number of lags to be included. The VAR in this case thus consists of 9 (the 8 of the first stage of the VECM in part 3.1 plus the unemployment rate) variables and their one-period lag. Contrary to the LSE approach, the dummy for the post-2008 crisis and the inflation of commodities could not be included as exogenous variables due to computation difficulties and we will therefore consider the possibility of the "price puzzle" in our analysis. Using a pure-sign approach – independent of the Cholesky ordering – and the restrictions mentioned before, Uhlig's (2005) algorithm was applied to construct the IRFs. The algorithm consists of first extracting the orthogonal innovations using Cholesky decomposition and derive the corresponding impulse response functions. Then, orthogonal impulse vectors for the shock of interest (shock in *Tbill3*) are randomly drawn and are multiplied to the IRFs. If the new responses are in accordance with the restrictions they are kept, otherwise discarded. The last step is repeated an arbitrary number of times so that it results in a set of approximately 900 potential IRFs, consistent with the restrictions imposed. Since this method does not yield a unique impulse response with its confidence interval, but a set of impulse responses, only the median will be depicted with a lower and upper bound representing the boundaries of the set estimated. It appeared that restricting inflation did not alter the responses of the other variables from the unrestricted case; we will thus focus on the most intuitive case, in which inflation is restricted. The resulting IRFs are presented in Figure 5. As noticed in part 3, all variables – except unemployment rate – (in terms of growth rates) tend to converge quickly to their initial rate. Hence, a contractionary shock does not seem to have a significant long-term impact on the growth of the variables. The potentially longer persistence of the response of unemployment rate might come from its own persistence and the possibly mistaken assumption of stationarity. The impulse response of output growth, the only unrestricted one, appear to be mostly negative and to go back to its initial level after slightly more than a year. Nonetheless, the magnitude of the response seems to be relatively small (a maximum impact of -0.07%). These results are in line with those found in part 3, even if in this case the interest rate is just shocked once and can vary afterwards. The set of impulse responses of GDP seems to include some positive responses as well, which could explain why in part 3 the effect of an increase in the short-term rate actually increased both GDP and its growth rate after a few months. The same IRFs with only 20-period horizons in which the 'short-term' effects are more apparent can be found in the appendix.

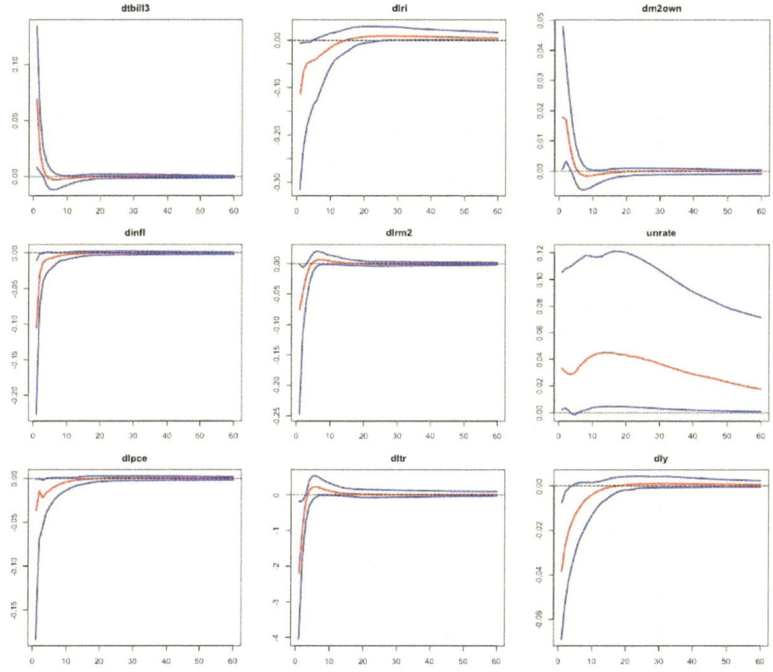

Figure 5: IRFs from a monetary policy shock

5 Conclusion

This paper aimed at evaluating the effects of monetary policy on GDP in the U.S. from 1983 to 2017. Throughout the paper we considered monetary policy shocks as a shock in the three-month Treasury Bill interest rate – a proxy for the Federal Funds rate. Two different approaches were used, in which 9 main macroeconomic variables were considered.

The first approach consisted in creating a system of equations based on a vector error correction model, hence focusing on the long-run relationships of the variables. This method enabled us to simulate a monetary policy in which the interest rate would have stayed around 4% at the end of 2007. We compared this simulation with forecasts using actual data for the exogenous

variables – baseline simulation. They were dynamically estimated, so that we could capture long-run effects of the interest rate as well. It turned out that increasing interest rates decreased GDP and its growth rate for about a year, but then after adjustments to the new high interest rate, GDP increased again and its level in the scenario actually slightly exceeded that of the baseline. Yet, the two simulations did not significantly differ from the actual series, and the effects of the interest rate did not appear to be substantial.
The second approach consisted of evaluating a Structural VAR on the stationary transformations of the data, which, contrary to the first method, focuses on contemporaneous relationships between the variables. Evaluating impulse response functions to a shock in the interest rate required an identification scheme. We used a pure-sign restriction approach, which has proved to be successful in investigating monetary policies in the past (Faust (1998), Uhlig (2005)). The method consists in imposing sign restrictions on the responses of the variables to the shock investigated, except on the one of interest. We based our restrictions on economic theory and on influential papers, such as those mentioned just before. While the responses vary between the variables, they all – except unemployment rate – seem to converge to their initial growth rate after approximately a year. The response of GDP seems to be in accordance with what we found in the first part, since it appears to be influenced with a small magnitude by the interest rate, and can as well slightly increase, as we have seen in the simulations.

The two approaches however significantly depend on the specifications and the data included. It is very likely that we are omitting variables and especially lacking information. There is probably, as mentioned in the 'price puzzle' example, an asymmetry between the information known by the policy makers and the information we retrieved from those few series. Moreover, the presence of structural breaks might as well significantly influence the relationship between the variables and it would be of further interest to investigate the models' stability between various subsamples.

A Appendix

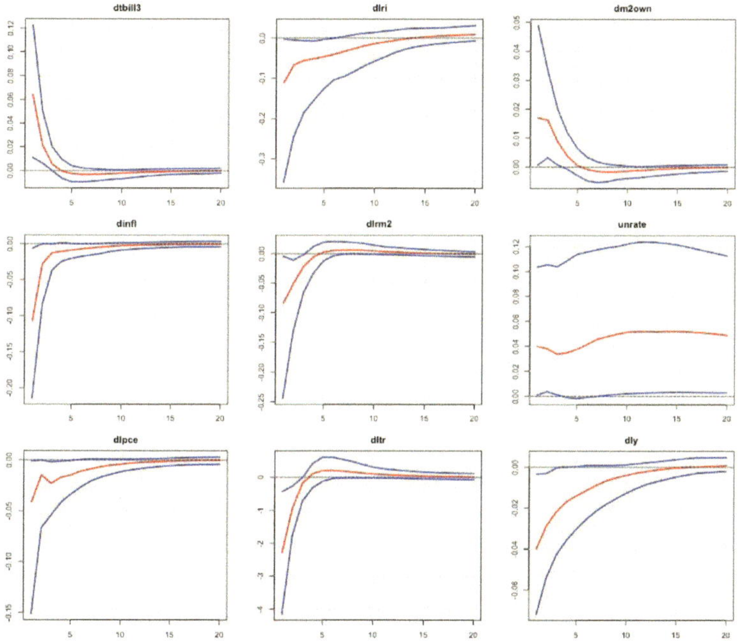

Figure 6: IRFs from a monetary policy shock

References

D'Amico, S., & King, T. B. (2015). What does anticipated monetary policy do?

Boivin, J., Kiley, M. T., Mishkin, F. S. (2010). How has the monetary transmission mechanism evolved over time? (No. w15879). National Bureau of Economic Research.

Castelnuovo, E. (2012). Monetary Policy Neutrality? Sign Restrictions Go to Monte Carlo (No. 0151). Dipartimento di Scienze Economiche" Marco Fanno".

Christians, L. J., Eichenbaum, M., & Evans, C. (1996). The Effects of Monetary Policy stocks. Evidence form the Flow Finds. The Review of Economic and statistics, 7811, 16-34.

Christiano, L. J., Eichenbaum, M., & Evans, C. L. (1997). Monetary policy shocks: what have we learned and to what end? (No. w6400). National bureau of economic research.

Faust, J. (1998, December). The robustness of identified VAR conclusions about money. In Carnegie-Rochester Conference Series on Public Policy (Vol. 49, pp. 207-244). North-Holland.

Favero, C. A. (2001). Applied macroeconometrics. Oxford University Press on Demand.

Gobbi, A., Willems, T. (2011). Identifying US monetary policy shocks through sign restrictions in dollarized countries.

Leeper, E. M., Sims, C. A., Zha, T., Hall, R. E., & Bernanke, B. S. (1996). What does monetary policy do?. Brookings papers on economic activity, 1996(2), 1-78.

Pfaff, B. (2008). Analysis of integrated and cointegrated time series with R. Springer Science Business Media.

Ramey, V. A. (2016). Macroeconomic shocks and their propagation. Handbook of Macroeconomics, 2, 71-162.

Romer, C. D., Romer, D. H. (1997). Identification and the narrative approach: A reply to Leeper. Journal of Monetary Economics, 40(3), 659-665.

Sims, C. A. (1992). Interpreting the macroeconomic time series facts: The effects of monetary policy. European economic review, 36(5), 975-1000.

Sims, Christopher A. and Tao Zha, 2006, "Does monetary policy generate recessions?" Macroeconomic Dynamics, Vol. 10, pp. 231–72.

Uhlig, H. (2005). What are the effects of monetary policy on output? Results from an agnostic identification procedure. Journal of Monetary Economics, 52(2), 381-419.